THE INTERNATIONAL REGULA
COLLISIONS AT

WITH AMENDMENTS ADOPTED FROM NOVEMBER 1995

Morgan Technical Books Limited
P.O. Box 5
Wotton-under-Edge
Glos. GL12 8HB
(01454) 261299

CONTENTS

PART A. GENERAL

RULE 1

Application

(*a*) These Rules shall apply to all vessels upon the high seas and in all waters connected therewith navigable by seagoing vessels.

(*b*) Nothing in these Rules shall interfere with the operation of special rules made by an appropriate authority for roadsteads, harbours, rivers, lakes or inland waterways connected with the high seas and navigable by seagoing vessels. Such special rules shall conform as closely as possible to these Rules.

(*c*) Nothing in these Rules shall interfere with the operation of any special rules made by the Government of any State with respect to additional station or signal lights, shapes or whistle signals for ships of war and vessels proceeding under convoy, or with respect to additional station or signal lights or shapes for fishing vessels engaged in fishing as a fleet. These additional station or signal lights, shapes or whistle signals shall, so far as possible, be such that they cannot be mistaken for any light, shape or signal authorized elsewhere under these Rules.

(*d*) Traffic separation schemes may be adopted by the Organization for the purpose of these Rules.

(*e*) Whenever the Government concerned shall have determined that a vessel of special construction or purpose cannot comply fully with the provisions of any of these Rules with respect to the number, position, range or arc of visibility of lights or shapes as well as to the disposition and characteristics of sound-signalling appliances, such vessel shall comply with such other provisions in regard to the number, position, range or arc of visibility of lights or shapes, as well as to the disposition and characteristics of sound-signalling appliances, as her Government shall have determined to be the closest possible compliance with these Rules in respect of that vessel.

RULE 2

Responsibility

(*a*) Nothing in these Rules shall exonerate any vessel, or the owner, master or crew thereof, from the consequences of any neglect to comply with these Rules or of the neglect of any precaution which may be required by the ordinary practice of seamen, or by the special circumstances of the case.

(*b*) In construing and complying with these Rules due regard shall be had to all dangers of navigation and collision and to any special circumstances, including the limitations of the vessels involved, which may make a departure from these Rules necessary to avoid immediate danger.

Rule 3

General definitions

For the purpose of these Rules, except where the context otherwise requires:

(a) The word "vessel" includes every description of water craft, including non-displacement craft and seaplanes, used or capable of being used as a means of transportation on water.

(b) The term "power-driven vessel" means any vessel propelled by machinery.

(c) The term "sailing vessel" means any vessel under sail provided that propelling machinery, if fitted, is not being used.

(d) The term "vessel engaged in fishing" means any vessel fishing with nets, lines, trawls or other fishing apparatus which restrict manoeuvrability, but does not include a vessel fishing with trolling lines or other fishing apparatus which do not restrict manoeuvrability.

(e) The word "seaplane" includes any aircraft designed to manoeuvre on the water.

(f) The term "vessel not under command" means a vessel which through some exceptional circumstance is unable to manoeuvre as required by these Rules and is therefore unable to keep out of the way of another vessel.

(g) The term "vessel restricted in her ability to manoeuvre" means a vessel which from the nature of her work is restricted in her ability to manoeuvre as required by these Rules and is therefore unable to keep out of the way of another vessel.

The term 'vessels restricted in their ability to manoeuvre' shall include but not be limited to:

(i) a vessel engaged in laying, servicing or picking up a navigation mark, submarine cable or pipeline;

(ii) a vessel engaged in dredging, surveying or underwater operations;

(iii) a vessel engaged in replenishment or transferring persons, provisions or cargo while underway;

(iv) a vessel engaged in the launching or recovery of aircraft;

(v) a vessel engaged in mineclearance operations;

(vi) a vessel engaged in a towing operation such as severely restricts the towing vessel and her tow in their ability to deviate from their course.

(h) The term "vessel constrained by her draught" means a power-driven vessel which, because of her draught in relation to the available depth and width of navigable water, is severely restricted in her ability to deviate from the course she is following.

(i) The word "underway" means that a vessel is not at anchor, or made fast to the shore, or aground.

(*j*) The words "length" and "breadth" of a vessel mean her length overall and greatest breadth.

(*k*) Vessels shall be deemed to be in sight of one another only when one can be observed visually from the other.

(*l*) The term "restricted visibility" means any condition in which visibility is restricted by fog, mist, falling snow, heavy rainstorms, sandstorms or any other similar causes.

PART B. STEERING AND SAILING RULES

Section I. Conduct of vessels in any condition of visibility

RULE 4

Application

Rules in this Section apply in any condition of visibility.

Rule 5

Look-out

Every vessel shall at all times maintain a proper look-out by sight and hearing as well as by all available means appropriate in the prevailing circumstance and conditions so as to make a full appraisal of the situation and of the risk of collision.

Rule 6

Safe speed

Every vessel shall at all times proceed at a safe speed so that she can take proper and effective action to avoid collision and be stopped within a distance appropriate to the prevailing circumstances and conditions.

In determining a safe speed the following factors shall be among those taken into account:

(*a*) By all vessels:

    (i) the state of visibility;

    (ii) the traffic density including concentrations of fishing vessels or any other vessels;

(iii) the manoeuvrability of the vessel with special reference to stopping distance and turning ability in the prevailing conditions;

(iv) at night the presence of background light such as from shore lights or from back scatter of her own lights;

(v) the state of wind, sea and current, and the proximity of navigational hazards;

(vi) the draught in relation to the available depth of water.

(b) Additionally, by vessels with operational radar:

(i) the characteristics, efficiency and limitations of the radar equipment;

(ii) any constraints imposed by the radar range scale in use;

(iii) the effect on radar detection of the sea state, weather and other sources of interference;

(iv) the possibility that small vessels, ice and other floating objects may not be detected by radar at an adequate range;

(v) the number, location and movement of vessels detected by radar;

(vi) the more exact assessment of the visibility that may be possible when radar is used to determine the range of vessels or other objects in the vicinity.

Rule 7
Risk of collision

(*a*) Every vessel shall use all available means appropriate to the prevailing circumstances and conditions to determine if risk of collision exists. If there is any doubt such risk shall be deemed to exist.

(*b*) Proper use shall be made of radar equipment if fitted and operational, including long-range scanning to obtain early warning of risk of collision and radar plotting or equivalent systematic observation of detected objects.

(*c*) Assumptions shall not be made on the basis of scanty information, especially scanty radar information.

(*d*) In determining if risk of collision exists the following considerations shall be among those taken into account:

  (i) such risk shall be deemed to exist if the compass bearing of an approaching vessel does not appreciably change;

  (ii) such risk may sometimes exist even when an appreciable bearing change is evident, particularly when approaching a very large vessel or a tow or when approaching a vessel at close range.

Rule 8
Action to avoid collision

(*a*) Any action taken to avoid collision shall, if the circumstances of the case admit, be positive, made in ample time and with due regard to the observance of good seamanship.

(*b*) Any alteration of course and/or speed to avoid collision shall, if the circumstances of the case admit, be large enough to be readily apparent to another vessel observing visually or by radar; a succession of small alterations of course and/or speed should be avoided.

(*c*) If there is sufficient sea room, alteration of course alone may be the most effective action to avoid a close-quarters situation provided that it is made in good time, is substantial and does not result in another close-quarters situation.

(*d*) Action taken to avoid collision with another vessel shall be such as to result in passing at a safe distance. The effectiveness of the action shall be carefully checked until the other vessel is finally past and clear.

(*e*) If necessary to avoid collision or allow more time to assess the situation, a vessel shall slacken her speed or take all way off by stopping or reversing her means of propulsion.

(*f*) (i) A vessel which, by any of these Rules, is required not to impede the passage or safe passage of another vessel shall, when required by the circumstances of the case, take early action to allow sufficient sea room for the safe passage of the other vessel.

(ii) A vessel required not to impede the passage or safe passage of another vessel is not relieved of this obligation if approaching the other vessel so as to involve risk of collision and shall, when taking action, have full regard to the action which may be required by the Rules of this part.

(iii) A vessel the passage of which is not to be impeded remains fully obliged to comply with the Rules of this part when the two vessels are approaching one another so as to involve risk of collision.

RULE 9

Narrow channels

(*a*) A vessel proceeding along the course of a narrow channel or fairway shall keep as near to the outer limit of the channel or fairway which lies on her starboard side as is safe and practicable.

(*b*) A vessel of less than 20 metres in length or a sailing vessel shall not impede the passage of a vessel which can safely navigate only within a narrow channel or fairway.

(*c*) A vessel engaged in fishing shall not impede the passage of any other vessel navigating within a narrow channel or fairway.

(*d*) A vessel shall not cross a narrow channel or fairway if such crossing impedes the passage of a vessel which can safely navigate only within such channel or fairway. The latter vessel may use the sound signal prescribed in Rule 34 (*d*) if in doubt as to the intention of the crossing vessel.

(*e*) (i) In a narrow channel or fairway when overtaking can take place only if the vessel to be overtaken has to take action to permit safe passing, the vessel intending to overtake shall indicate her intention by sounding the appropriate signal prescribed in Rule 34 (*c*) (i). The vessel to be overtaken shall, if in agreement, sound the appropriate signal prescribed in Rule 34 (*c*) (ii) and take steps to permit safe passing. If in doubt she may sound the signals prescribed in Rule 34 (*d*).

(ii) This Rule does not relieve the overtaking vessel of her obligation under Rule 13.

(*f*) A vessel nearing a bend or an area of a narrow channel or fairway where other vessels may be obscured by an intervening obstruction shall navigate with particular alertness and caution and shall sound the appropriate signal prescribed in Rule 34 (*e*).

(*g*) Any vessel shall, if the circumstances of the case admit, avoid anchoring in a narrow channel.

RULE 10

Traffic separation schemes

(*a*) This Rule applies to traffic separation schemes adopted by the Organization and does not relieve any vessel of her obligation under any other Rule.

(*b*) A vessel using a traffic separation scheme shall:

(i) proceed in the appropriate traffic lane in the general direction of traffic flow for that lane;

(ii) so far as practicable keep clear of a traffic separation line or separation zone;

(iii) normally join or leave a traffic lane at the termination of the lane, but when joining or leaving from either side shall do so at as small an angle to the general direction of traffic flow as practicable.

(*c*) A vessel shall, so far as practicable, avoid crossing traffic lanes but if obliged to do so shall cross on a heading as nearly as practicable at right angles to the general direction of traffic flow.

(d) (i) A vessel shall not use an inshore traffic zone when she can safely use the appropriate traffic lane within the adjacent traffic separation scheme. However, vessels of less than 20 metres in length, sailing vessels and vessels engaged in fishing may use the inshore traffic zone.

(ii) Notwithstanding subparagraph *(d)* (i), a vessel may use an inshore traffic zone when *en route* to or from a port, offshore installation or structure, pilot station or any other place situated within the inshore traffic zone, or to avoid immediate danger.

(e) A vessel other than a crossing vessel or a vessel joining or leaving a lane shall not normally enter a separation zone or cross a separation line except:

(i) in cases of emergency to avoid immediate danger;

(ii) to engage in fishing within a separation zone.

(f) A vessel navigating in areas near the terminations of traffic separation schemes shall do so with particular caution.

(g) A vessel shall so far as practicable avoid anchoring in a traffic separation scheme or in areas near its terminations.

(h) A vessel not using a traffic separation scheme shall avoid it by as wide a margin as practicable.

(i) A vessel engaged in fishing shall not impede the passage of any vessel following a traffic lane.

(j) A vessel of less than 20 metres in length or a sailing vessel shall not impede the safe passage of a power-driven vessel following a traffic lane.

(k) A vessel restricted in her ability to manoeuvre when engaged in an operation for the maintenance of safety of navigation in a traffic separation scheme is exempted from complying with this Rule to the extent necessary to carry out the operation.

(l) A vessel restricted in her ability to manoeuvre when engaged in an operation for the laying, servicing or picking up of a submarine cable, within a traffic separation scheme, is exempted from complying with this Rule to the extent necessary to carry out the operation.

Section II. Conduct of vessels in sight of one another

RULE 11

Application

Rules in this Section apply to vessels in sight of one another.

RULE 12

Sailing vessels

(*a*) When two sailing vessels are approaching one another, so as to involve risk of collision, one of them shall keep out of the way of the other as follows:

    (i) when each has the wind on a different side, the vessel which has the wind on the port side shall keep out of the way of the other;

    (ii) when both have the wind on the same side, the vessel which is to windward shall keep out of the way of the vessel which is to leeward;

    (iii) if a vessel with the wind on the port side sees a vessel to windward and cannot determine with certainty whether the other vessel has the wind on the port or on the starboard side, she shall keep out of the way of the other.

(*b*) For the purposes of this Rule the windward side shall be deemed to be the side opposite to that on which the mainsail is carried or, in the case of a square-rigged vessel, the side opposite to that on which the largest fore-and-aft sail is carried.

Rule 13

Overtaking

(*a*) Notwithstanding anything contained in the Rules of Part B, Sections I and II any vessel overtaking any other shall keep out of the way of the vessel being overtaken.

(*b*) A vessel shall be deemed to be overtaking when coming up with another vessel from a direction more than 22·5 degrees abaft her beam, that is, in such a position with reference to the vessel she is overtaking, that at night she would be able to see only the sternlight of that vessel but neither of her sidelights.

(*c*) When a vessel is in any doubt as to whether she is overtaking another, she shall assume that this is the case and act accordingly.

(*d*) Any subsequent alteration of the bearing between the two vessels shall not make the overtaking vessel a crossing vessel within the meaning of these Rules or relieve her of the duty of keeping clear of the overtaken vessel until she is finally past and clear.

Rule 14

Head-on situation

(*a*) When two power-driven vessels are meeting on reciprocal or nearly reciprocal courses so as to involve risk of collision each shall alter her course to starboard so that each shall pass on the port side of the other.

(*b*) Such a situation shall be deemed to exist when a vessel sees the other ahead or nearly ahead and by night she could see the masthead lights of the other in a line or nearly in a line and/or both sidelights and by day she observes the corresponding aspect of the other vessel.

(*c*) When a vessel is in any doubt as to whether such a situation exists she shall assume that it does exist and act accordingly.

RULE 15

Crossing situation

When two power-driven vessels are crossing so as to involve risk of collision, the vessel which has the other on her own starboard side shall keep out of the way and shall, if the circumstances of the case admit, avoid crossing ahead of the other vessel.

RULE 16

Action by give-way vessel

Every vessel which is directed to keep out of the way of another vessel shall, so far as possible, take early and substantial action to keep well clear.

RULE 17

Action by stand-on vessel

(*a*) (i) Where one of two vessels is to keep out of the way the other shall keep her course and speed.

(ii) The latter vessel may however take action to avoid collision by her manoeuvre alone, as soon as it becomes apparent to her that the vessel required to keep out of the way is not taking appropriate action in compliance with these Rules.

(b) When, from any cause, the vessel required to keep her course and speed finds herself so close that collision cannot be avoided by the action of the give-way vessel alone, she shall take such action as will best aid to avoid collision.

(c) A power-driven vessel which takes action in a crossing situation in accordance with sub-paragraph (a) (ii) of this Rule to avoid collision with another power-driven vessel shall, if the circumstances of the case admit, not alter course to port for a vessel on her own port side.

(d) This Rule does not relieve the give-way vessel of her obligation to keep out of the way.

RULE 18

Responsibilities between vessels

Except where Rules 9, 10 and 13 otherwise require:

(a) A power-driven vessel underway shall keep out of the way of:

     (i) a vessel not under command;

     (ii) a vessel restricted in her ability to manoeuvre;

     (iii) a vessel engaged in fishing;

     (iv) a sailing vessel.

(*b*) A sailing vessel underway shall keep out of the way of:

     (i) a vessel not under command;

     (ii) a vessel restricted in her ability to manoeuvre;

     (iii) a vessel engaged in fishing.

(*c*) A vessel engaged in fishing when underway shall, so far as possible, keep out of the way of:

     (i) a vessel not under command;

     (ii) a vessel restricted in her ability to manoeuvre.

(*d*)   (i) Any vessel other than a vessel not under command or a vessel restricted in her ability to manoeuvre shall, if the circumstances of the case admit, avoid impeding the safe passage of a vessel constrained by her draught, exhibiting the signals in Rule 28.

     (ii) A vessel constrained by her draught shall navigate with particular caution having full regard to her special condition.

(*e*) A seaplane on the water shall, in general, keep well clear of all vessels and avoid impeding their navigation. In circumstances, however, where risk of collision exists, she shall comply with the Rules of this Part.

Section III. Conduct of vessels in restricted visibility

RULE 19

Conduct of vessels in restricted visibility

(*a*) This Rule applies to vessels not in sight of one another when navigating in or near an area of restricted visibility.

(*b*) Every vessel shall proceed at a safe speed adapted to the prevailing circumstances and conditions of restricted visibility. A power-driven vessel shall have her engines ready for immediate manoeuvre.

(*c*) Every vessel shall have due regard to the prevailing circumstances and conditions of restricted visibility when complying with the Rules of Section I of this Part.

(*d*) A vessel which detects by radar alone the presence of another vessel shall determine if a close-quarters situation is developing and/or risk of collision exists. If so, she shall take avoiding action in ample time, provided that when such action consists of an alteration of course, so far as possible the following shall be avoided:

(i) an alteration of course to port for a vessel forward of the beam, other than for a vessel being overtaken;

(ii) an alteration of course towards a vessel abeam or abaft the beam.

(*e*) Except where it has been determined that a risk of collision does not exist, every vessel which hears apparently forward of her beam the fog signal of another vessel, or which cannot avoid a close-quarters situation with another vessel forward of her beam, shall reduce her speed to the minimum at which she can be kept on her course. She shall if necessary take all her way off and in any event navigate with extreme caution until danger of collision is over.

PART C. LIGHTS AND SHAPES

RULE 20

Application

(*a*) Rules in this Part shall be complied with in all weathers.

(*b*) The Rules concerning lights shall be complied with from sunset to sunrise, and during such times no other lights shall be exhibited, except such lights as cannot be mistaken for the lights specified in these Rules or do not impair their visibility or distinctive character, or interfere with the keeping of a proper look-out.

(c) The lights prescribed by these Rules shall, if carried, also be exhibited from sunrise to sunset in restricted visibility and may be exhibited in all other circumstances when it is deemed necessary.

(d) The Rules concerning shapes shall be complied with by day.

(e) The lights and shapes specified in these Rules shall comply with the provisions of Annex I to these Regulations.

Rule 21

Definitions

(a) "Masthead light" means a white light placed over the fore and aft centre-line of the vessel showing an unbroken light over an arc of the horizon of 225 degrees and so fixed as to show the light from right ahead to 22·5 degrees abaft the beam on either side of the vessel.

(b) "Sidelights" means a green light on the starboard side and a red light on the port side each showing an unbroken light over an arc of the horizon of 112·5 degrees and so fixed as to show the light from right ahead to 22·5 degrees abaft the beam on its respective side. In a vessel of less than 20 metres in length the sidelights may be combined in one lantern carried on the fore and aft centreline of the vessel.

(c) "Sternlight" means a white light placed as nearly as practicable at the stern showing an unbroken light over an arc of the horizon of 135 degrees and so fixed as to show the light 67·5 degrees from right aft on each side of the vessel.

(*d*) "Towing light" means a yellow light having the same characteristics as the "sternlight" defined in paragraph (*c*) of this Rule.

(*e*) "All round light" means a light showing an unbroken light over an arc of the horizon of 360 degrees.

(*f*) "Flashing light" means a light flashing at regular intervals at a frequency of 120 flashes or more per minute.

RULE 22

Visibility of lights

The lights prescribed in these Rules shall have an intensity as specified in Section 8 of Annex I to these Regulations so as to be visible at the following minimum ranges:

(*a*) In vessels of 50 metres or more in length:

—a masthead light, 6 miles;

—a sidelight, 3 miles;

—a sternlight, 3 miles;

—a towing light, 3 miles;

—a white, red, green or yellow all-round light, 3 miles.

(b) In vessels of 12 metres or more in length but less than 50 metres in length:

    —a masthead light, 5 miles; except that where the length of the vessel is less than 20 metres, 3 miles;

    —a sidelight, 2 miles;

    —a sternlight, 2 miles;

    —a towing light, 2 miles;

    —a white, red, green or yellow all-round light, 2 miles.

(c) In vessels of less than 12 metres in length:

    —a masthead light, 2 miles;

    —a sidelight, 1 mile;

    —a sternlight, 2 miles;

    —a towing light, 2 miles;

    —a white, red, green or yellow all-round light, 2 miles.

(d) –In inconspicuous, partly submerged vessels or objects being towed:

    –a white all-round light, 3 miles.

Rule 23

Power-driven vessels underway

(a) A power-driven vessel underway shall exhibit:

  (i) a masthead light forward;

  (ii) a second masthead light abaft of and higher than the forward one; except that a vessel of less than 50 metres in length shall not be obliged to exhibit such light but may do so;

  (iii) sidelights;

  (iv) a sternlight.

(b) An air-cushion vessel when operating in the non-displacement mode shall, in addition to the lights prescribed in paragraph *(a)* of this Rule, exhibit an all-round flashing yellow light.

(c) (i) A power-driven vessel of less than 12 metres in length may in lieu of the lights prescribed in paragraph *(a)* of this Rule exhibit an all-round white light and sidelights;

  (ii) a power-driven vessel of less than 7 metres in length whose maximum speed does not exceed 7 knots may in lieu of the lights prescribed in paragraph *(a)* of this Rule exhibit an all-round white light and shall, if practicable, also exhibit sidelights;

(iii) the masthead light or all-round white light on a power-driven vessel of less than 12 metres in length may be displaced from the fore and aft centreline of the vessel if centreline fitting is not practicable, provided that the sidelights are combined in one lantern which shall be carried on the fore and aft centreline of the vessel or located as nearly as practicable in the same fore and aft line as the masthead light or the all-round white light.

RULE 24

Towing and pushing

(*a*) A power-driven vessel when towing shall exhibit:

  (i) instead of the light prescribed in Rule 23 *(a)* (i) or *(a)* (ii), two masthead lights in a vertical line. When the length of the tow, measuring from the stern of the towing vessel to the after end of the tow exceeds 200 metres, three such lights in a vertical line;

 (ii) sidelights;

(iii) a sternlight;

(iv) a towing light in a vertical line above the sternlight;

 (v) when the length of the tow exceeds 200 metres, a diamond shape where it can best be seen.

(*b*) When a pushing vessel and a vessel being pushed ahead are rigidly connected in a composite unit they shall be regarded as a power-driven vessel and exhibit the lights prescribed in Rule 23.

(*c*) A power-driven vessel when pushing ahead or towing alongside, except in the case of a composite unit shall exhibit:

(i) instead of the light prescribed in Rule 23 *(a)* (i) or *(a)* (ii), two masthead lights in a vertical line;

(ii) sidelights;

(iii) a sternlight.

(*d*) A power-driven vessel to which paragraphs *(a)* or *(c)* of this Rule apply shall also comply with Rule 23 *(a)* (ii).

(*e*) A vessel or object being towed, other than those mentioned in paragraph *(g)* of this Rule, shall exhibit:

(i)sidelights,

(ii)a sternlight;

(iii)when the length of the tow exceeds 200 metres, a diamond shape where it can best be seen.

(*f*) Provided that any number of vessels being towed alongside or pushed in a group shall be lighted as one vessel,

    (i)a vessel being pushed ahead, not being part of a composite unit, shall exhibit at the forward end, sidelights;

    (ii)a vessel being towed alongside shall exhibit a sternlight and at the forward end, sidelights.

(*g*) An inconspicuous, partly submerged vessel or object, or combination of such vessels or objects being towed, shall exhibit:

    (i)if it is less than 25 metres in breadth, one all-round white light at or near the forward end and one at or near the after end except that dracones need not exhibit a light at or near the forward end.

    (ii)if it is 25 metres or more in breadth, two additional all-round white lights at or near the extremities of its breadth;

    (iii)if it exceeds 100 metres in length, additional all-round white lights between the lights prescribed in sub-paragraphs (i) and (ii) so that the distance between the lights shall not exceed 100 metres;

(iv) a diamond shape at or near the aftermost extremity of the last vessel or object being towed and if the length of the tow exceeds 200 metres an additional diamond shape where it can best be seen and located as far forward as is practicable.

(h) Where from any sufficient cause it is impracticable for a vessel or object being towed to exhibit the lights or shapes prescribed in paragraph *(e)* or *(g)* of this Rule, all possible measures shall be taken to light the vessel or object towed or at least to indicate the presence of such vessel or object.

(i) Where from any sufficient cause it is impracticable for a vessel not normally engaged in towing operations to display the lights prescribed in paragraph *(a)* or *(c)* of this Rule, such vessel shall not be required to exhibit those lights when engaged in towing another vessel in distress or otherwise in need of assistance. All possible measures shall be taken to indicate the nature of the relationship between the towing vessel and the vessel being towed as authorized by Rule 36, in particular by illuminating the towline.

Rule 25

Sailing vessels underway and vessels under oars

(*a*) A sailing vessel underway shall exhibit:

  (i) sidelights;

  (ii) a sternlight.

(*b*) In a sailing vessel of less than 20 metres in length the lights prescribed in paragraph (*a*) of this Rule may be combined in one lantern carried at or near the top of the mast where it can best be seen.

(*c*) A sailing vessel underway may, in addition to the lights prescribed in paragraph (*a*) of this Rule, exhibit at or near the top of the mast, where they can best be seen, two all-round lights in a vertical line, the upper being red and the lower green, but these lights shall not be exhibited in conjunction with the combined lantern permitted by paragraph (*b*) of this Rule.

(*d*) (i) A sailing vessel of less than 7 metres in length shall, if practicable, exhibit the lights prescribed in paragraph (*a*) or (*b*) of this Rule, but if she does not, she shall have ready at hand an electric torch or lighted lantern showing a white light which shall be exhibited in sufficient time to prevent collision.

(ii) A vessel under oars may exhibit the lights prescribed in this Rule for sailing vessels, but if she does not, she shall have ready at hand an electric torch or lighted lantern showing a white light which shall be exhibited in sufficient time to prevent collision.

(*e*) A vessel proceeding under sail when also being propelled by machinery shall exhibit forward where it can best be seen a conical shape, apex downwards.

Rule 26

Fishing vessels

(*a*) A vessel engaged in fishing, whether underway or at anchor, shall exhibit only the lights and shapes prescribed in this Rule.

(*b*) A vessel when engaged in trawling, by which is meant the dragging through the water of a dredge net or other apparatus used as a fishing appliance, shall exhibit:

  (i) two all-round lights in a vertical line, the upper being green and the lower white, or a shape consisting of two cones with their apexes together in a vertical line one above the other;

  (ii) a masthead light abaft of and higher than the all-round green light; a vessel of less than 50 metres in length shall not be obliged to exhibit such a light but may do so;

  (iii) when making way through the water, in addition to the lights prescribed in this paragraph, sidelights and a sternlight.

(*c*) A vessel engaged in fishing, other than trawling, shall exhibit:

  (i) two all-round lights in a vertical line, the upper being red and the lower white, or a shape consisting of two cones with apexes together in a vertical line one above the other;

(ii) when there is outlying gear extending more than 150 metres horizontally from the vessel, an all-round white light or a cone apex upwards in the direction of the gear;

(iii) when making way through the water, in addition to the lights prescribed in this paragraph, sidelights and a sternlight.

(d) The additional signals described in Annex II to these regulations apply to a vessel engaged in fishing in close proximity to other vessels engaged in fishing.

(e) A vessel when not engaged in fishing shall not exhibit the lights or shapes prescribed in this Rule, but only those prescribed for a vessel of her length.

RULE 27

Vessels not under command or restricted in their ability to manoeuvre

(a) A vessel not under command shall exhibit:

(i) two all-round red lights in a vertical line where they can best be seen;

(ii) two balls or similar shapes in a vertical line where they can best be seen;

(iii) when making way through the water, in addition to the lights prescribed in this paragraph, sidelights and a sternlight.

(b) A vessel restricted in her ability to manoeuvre, except a vessel engaged in mineclearance operations, shall exhibit:

  (i)three all-round lights in a vertical line where they can best be seen. The highest and lowest of these lights shall be red and the middle light shall be white;

  (ii)three shapes in a vertical line where they can best be seen. The highest and lowest of these shapes shall be balls and the middle one a diamond;

  (iii)when making way through the water, a masthead light or lights, sidelights and a sternlight, in addition to the lights prescribed in sub-paragraph (i);

  (iv)when at anchor, in addition to the lights or shapes prescribed in sub-paragraphs (i) and (ii), the light, lights or shape prescribed in Rule 30.

(c) A power-driven vessel engaged in a towing operation such as severely restricts the towing vessel and her tow in their ability to deviate from their course shall, in addition to the lights or shapes prescribed in Rule 24 *(a)*, exhibit the lights or shapes prescribed in sub-paragraphs *(b)* (i) and (ii) of this Rule.

(d) A vessel engaged in dredging or underwater operations, when restricted in her ability to manoeuvre, shall exhibit the lights and shapes prescribed in sub-paragraph *(b)* (i), (ii) and (iii) of this Rule and shall in addition, when an obstruction exists, exhibit:

  (i)two all-round red lights or two balls in a vertical line to indicate the side on which the obstruction exists;

      (ii)two all-round green lights or two diamonds in a vertical line to indicate the side on which another vessel may pass;

      (iii)when at anchor the lights or shapes prescribed in this paragraph instead of the lights or shape prescribed in Rule 30.

(e) Whenever the size of a vessel engaged in diving operations makes it impracticable to exhibit all lights and shapes prescribed in paragraph *(d)* of this Rule, the following shall be exhibited:

      (i)three all-round lights in a vertical line where they can best be seen. The highest and lowest of these lights shall be red and the middle light shall be white;

      (ii)a rigid replica of the International Code Flag "A" not less than 1 metre in height. Measures shall be taken to ensure its all-round visibility.

(f) A vessel engaged in mineclearance operations shall in addition to the lights prescribed for a power-driven vessel in Rule 23 or to the lights or shape prescribed for a vessel at anchor in Rule 30 as appropriate, exhibit three all-round green lights or three balls. One of these lights or shapes shall be exhibited near the foremast head and one at each end of the fore yard. These lights or shapes indicate that it is dangerous for another vessel to approach within 1000 metres of the mineclearance vessel.

(g) Vessels of less than 12 metres in length, except those engaged in diving operations shall not be required to exhibit the lights and shapes prescribed in this Rule.

(*h*) The signals prescribed in this Rule are not signals of vessels in distress and requiring assistance. Such signals are contained in Annex IV to these Regulations.

RULE 28

Vessels constrained by their draught

A vessel constrained by her draught may, in addition to the lights prescribed for power-driven vessels in Rule 23, exhibit where they can best be seen three all-round red lights in a verticle line, or a cylinder.

RULE 29

Pilot vessels

(*a*) A vessel engaged on pilotage duty shall exhibit:

  (i) at or near the masthead, two all-round lights in a vertical line, the upper being white and the lower red;

 (ii) when underway, in addition, sidelights and a sternlight;

(iii) when at anchor, in addition to the lights prescribed in sub-paragraph(i), the light, lights or shape prescribed in Rule 30 for vessels at anchor.

(*b*) A pilot vessel when not engaged on pilotage duty shall exhibit the lights or shapes prescribed for a similar vessel of her length.

Rule 30

Anchored vessels and vessels aground

(*a*) A vessel at anchor shall exhibit where it can best be seen:

  (i) in the fore part, an all-round white light or one ball;

  (ii) at or near the stern and at a lower level than the light prescribed in sub-paragraph (i), an all-round white light.

(*b*) A vessel of less than 50 metres in length may exhibit an all-round white light where it can best be seen instead of the lights prescribed in paragraph (*a*) of this Rule.

(*c*) A vessel at anchor may, and a vessel of 100 metres and more in length shall, also use the available working or equivalent lights to illuminate her decks.

(*d*) A vessel aground shall exhibit the lights prescribed in paragraph (*a*) or (*b*) of this Rule and in addition, where they can best be seen:

  (i) two all-round red lights in a vertical line;

  (ii) three balls in a vertical line.

(*e*) A vessel of less than 7 metres in length, when at anchor, not in or near a narrow channel, fairway or anchorage, or where other vessels normally navigate, shall not be required to exhibit the lights or shape prescribed in paragraphs (*a*) and (*b*) of this Rule.

(*f*) A vessel of less than 12 metres in length, when aground, shall not be required to exhibit the lights or shapes prescribed in sub-paragraphs (*d*) (i) and (ii) of this Rule.

Rule 31

Seaplanes

Where it is impracticable for a seaplane to exhibit lights and shapes of the characteristics or in the positions prescribed in the Rules of this Part she shall exhibit lights and shapes as closely similar in characteristics and position as is possible.

PART D. SOUND AND LIGHT SIGNALS

Rule 32

Definitions

(*a*) The word "whistle" means any sound signalling appliance capable of producing the prescribed blasts and which complies with the specifications in Annex III to these Regulations.

(*b*) The term "short blast" means a blast of about one second's duration.

(*c*) The term "prolonged blast" means a blast of from four to six seconds' duration.

RULE 33

Equipment for sound signals

(*a*) A vessel of 12 metres or more in length shall be provided with a whistle and a bell and a vessel of 100 metres or more in length shall, in addition, be provided with a gong, the tone and sound of which cannot be confused with that of the bell. The whistle, bell and gong shall comply with the specifications in Annex III to these Regulations. The bell or gong or both may be replaced by other equipment having the same respective sound characteristics, provided that manual sounding of the prescribed signals shall always be possible.

(*b*) A vessel of less than 12 metres in length shall not be obliged to carry the sound signalling appliances prescribed in paragraph (*a*) of this Rule but if she does not, she shall be provided with some other means of making an efficient sound signal.

RULE 34

Manoeuvring and warning signals

(*a*) When vessels are in sight of one another, a power-driven vessel underway, when manoeuvring as authorized or required by these Rules, shall indicate that manoeuvre by the following signals on her whistle:

—one short blast to mean "I am altering my course to starboard";

—two short blasts to mean "I am altering my course to port";

—three short blasts to mean "I am operating astern propulsion".

(b) Any vessel may supplement the whistle signals prescribed in paragraph (a) of this Rule by light signals, repeated as appropriate, whilst the manoeuvre is being carried out:

(i) these light signals shall have the following significance:

—one flash to mean "I am altering my course to starboard";

—two flashes to mean "I am altering my course to port";

—three flashes to mean "I am operating astern propulsion";

(ii) the duration of each flash shall be about one second, the interval between flashes shall be about one second, and the interval between successive signals shall be not less than ten seconds;

(iii) the light used for this signal shall, if fitted, be an all-round white light, visible at a minimum range of 5 miles, and shall comply with the provisions of Annex I to these Regulations.

(c) When in sight of one another in a narrow channel or fairway:

(i) a vessel intending to overtake another shall in compliance with Rule 9 (e) (i) indicate her intention by the following signals on her whistle:

—two prolonged blasts followed by one short blast to mean "I intend to overtake you on your starboard side";

—two prolonged blasts followed by two short blasts to mean "I intend to overtake you on your port side".

(ii) the vessel about to be overtaken when acting in accordance with Rule 9 (*e*) (i) shall indicate her agreement by the following signal on her whistle:

one prolonged, one short, one prolonged and one short blast, in that order.

(*d*) When vessels in sight of one another are approaching each other and from any cause either vessel fails to understand the intentions or actions of the other, or is in doubt whether sufficient action is being taken by the other to avoid collision, the vessel in doubt shall immediately indicate such doubt by giving at least five short and rapid blasts on the whistle. Such signal may be supplemented by a light signal of at least five short and rapid flashes.

(*e*) A vessel nearing a bend or an area of a channel or fairway where other vessels may be obscured by an intervening obstruction shall sound one prolonged blast. Such signal shall be answered with a prolonged blast by any approaching vessel that may be within hearing around the bend or behind the intervening obstruction.

(*f*) If whistles are fitted on a vessel at a distance apart of more than 100 metres, one whistle only shall be used for giving manoeuvring and warning signals.

Rule 35

Sound signals in restricted visibility

In or near an area of restricted visibility, whether by day or night, the signals prescribed in this Rule shall be used as follows:

(a) A power-driven vessel making way through the water shall sound at intervals of not more than 2 minutes one prolonged blast.

(b) A power-driven vessel underway but stopped and making no way through the water shall sound at intervals of not more than 2 minutes two prolonged blasts in succession with an interval of about 2 seconds between them.

(c) A vessel not under command, a vessel restricted in her ability to manoeuvre, a vessel constrained by her draught, a sailing vessel, a vessel engaged in fishing and a vessel engaged in towing or pushing another vessel shall, instead of the signals prescribed in paragraphs (a) or (b) of this Rule, sound at intervals of not more than 2 minutes three blasts in succession, namely one prolonged followed by two short blasts.

(d) A vessel engaged in fishing, when at anchor, and a vessel restricted in her ability to manoeuvre when carrying out her work at anchor, shall instead of the signals prescribed in paragraph (g) of this Rule sound the signal prescribed in paragraph (c) of this Rule.

(e) A vessel towed or if more than one vessel is towed the last vessel of the tow, if manned, shall at intervals of not more than 2 minutes sound four blasts in succession, namely one prolonged followed by three short blasts. When practicable, this signal shall be made immediately after the signal made by the towing vessel.

(f) When a pushing vessel and a vessel being pushed ahead are rigidly connected in a composite unit they shall be regarded as a power-driven vessel and shall give the signals prescribed in paragraphs *(a)* or *(b)* of this Rule.

(g) A vessel at anchor shall at intervals of not more than one minute ring the bell rapidly for about 5 seconds. In a vessel of 100 metres or more in length the bell shall be sounded in the forepart of the vessel and immediately after the ringing of the bell the gong shall be sounded rapidly for about 5 seconds in the after part of the vessel. A vessel at anchor may in addition sound three blasts in succession, namely one short, one prolonged and one short blast, to give warning of her position and of the possibility of collision to an approaching vessel.

(h) A vessel aground shall give the bell signal and if required the gong signal prescribed in paragraph *(g)* of this Rule and shall, in addition, give three separate and distinct strokes on the bell immediately before and after the rapid ringing of the bell. A vessel aground may in addition sound an appropriate whistle signal.

(*i*) A vessel of less than 12 metres in length shall not be obliged to give the above-mentioned signals but, if she does not, shall make some other efficient sound signal at intervals of not more than 2 minutes.

(*j*) A pilot vessel when engaged on pilotage duty may in addition to the signals prescribed in paragraphs (*a*), (*b*) or (*f*) of this Rule sound an identity signal consisting of four short blasts.

RULE 36

Signals to attract attention

If necessary to attract the attention of another vessel any vessel may make light or sound signals that cannot be mistaken for any signal authorized elsewhere in these Rules, or may direct the beam of her searchlight in the direction of the danger, in such a way as not to embarrass any vessel. Any light to attract the attention of another vessel shall be such that it cannot be mistaken for any aid to navigation. For the purpose of this Rule the use of high intensity intermittent or revolving lights, such as strobe lights, shall be avoided.

RULE 37

Distress signals

When a vessel is in distress and requires assistance she shall use or exhibit the signals described in Annex IV to these Regulations.

PART E. EXEMPTIONS

RULE 38

Exemptions

Any vessel (or class of vessels) provided that she complies with the requirements of the International Regulations for Preventing Collisions at Sea, 1960, the keel of which is laid or which is at a corresponding stage of construction before the entry into force of these Regulations may be exempted from compliance therewith as follows:

(*a*) The installation of lights with ranges prescribed in Rule 22, until four years after the date of entry into force of these Regulations.

(*b*) The installation of lights with colour specifications as prescribed in Section 7 of Annex I to these Regulations, until four years after the date of entry into force of these Regulations.

(*c*) The repositioning of lights as a result of conversion from Imperial to metric units and rounding off measurement figures, permanent exemption.

(*d*) (i) The repositioning of masthead lights on vessels of less than 150 metres in length, resulting from the prescriptions of Section 3 *(a)* of Annex I to these Regulations, permanent exemption.

(ii) The repositioning of masthead lights on vessels of 150 metres or more in length, resulting from the prescriptions of Section 3 (*a*) of Annex I to these Regulations, until nine years after the date of entry into force of these Regulations.

(*e*) The repositioning of masthead lights resulting from the prescriptions of Section 2 (*b*) of Annex I to these Regulations until nine years after the date of entry into force of these Regulations.

(*f*) The repositioning of sidelights resulting from the prescriptions of Sections 2 (*g*) and 3 (*b*) of Annex I to these Regulations until nine years after the date of entry into force of these Regulations.

(*g*) The requirements for sound signal appliances prescribed in Annex III to these Regulations until nine years after the date of entry into force of these Regulations.

(*h*) The repositioning of all-round lights resulting from the prescription of Section 9 (*b*) of Annex I to these Regulations, permanent exemption.

ANNEX I

Positioning and technical details of lights and shapes

1. *Definition*

The term "height above the hull" means height above the uppermost continuous deck. This height shall be measured from the position vertically beneath the location of the light.

2. *Vertical positioning and spacing of lights*

(a) On a power-driven vessel of 20 metres or more in length the masthead lights shall be placed as follows:

   (i) the forward masthead light, or if only one masthead light is carried, then that light, at a height above the hull of not less than 6 metres, and, if the breadth of the vessel exceeds 6 metres, then at a height above the hull not less than such breadth, so however that the light need not be placed at a greater height above the hull than 12 metres;

   (ii) when two masthead lights are carried the after one shall be at least 4.5 metres vertically higher than the forward one.

(b) The vertical separation of masthead lights of power-driven vessels shall be such that in all normal conditions of trim the after light will be seen over and separate from the forward light at a distance of 1,000 metres from the stem when viewed from sea level.

(c) The masthead light of a power-driven vessel of 12 metres but less than 20 metres in length shall be placed at a height above the gunwale of not less than 2·5 metres.

(d) A power-driven vessel of less than 12 metres in length may carry the uppermost light at a height of less than 2.5 metres above the gunwale. When however a masthead light is carried in addition to sidelights and a sternlight or the all-round light prescribed in Rule 23*(c)*(i) is carried in addition to sidelights, then such masthead light or all-round light shall be carried at least 1 metre higher than the sidelights.

(e) One of the two or three masthead lights prescribed for a power-driven vessel when engaged in towing or pushing another vessel shall be placed in the same position as either the forward masthead light or the after masthead light; provided that, if carried on the aftermast, the lowest after masthead light shall be at least 4·5 metres vertically higher than the forward masthead light.

(f) (i) The masthead light or lights prescribed in Rule 23 *(a)* shall be so placed as to be above and clear of all other lights and obstructions except as described in sub-paragraph (ii).

(ii) When it is impracticable to carry the all-round lights prescribed by Rule 27 *(b)*(i) or Rule 28 below the masthead lights, they may be carried above the after masthead light(s) or vertically in between the forward masthead light(s) and after masthead light(s), provided that in the latter case the requirement of Section 3 *(c)* of this Annex shall be complied with.

(g) The sidelights of a power-driven vessel shall be placed at a height above the hull not greater than three-quarters of that of the forward masthead light. They shall not be so low as to be interfered with by deck lights.

(h) The sidelights, if in a combined lantern and carried on a power-driven vessel of less than 20 metres in length, shall be placed not less than 1 metre below the masthead light.

(i) When the Rules prescribe two or three lights to be carried in a vertical line, they shall be spaced as follows:

   (i) on a vessel of 20 metres in length or more such lights shall be spaced not less than 2 metres apart, and the lowest of these lights shall, except where a towing light is required, be placed at a height of not less than 4 metres above the hull;

   (ii) on a vessel of less than 20 metres in length such lights shall be spaced not less than 1 metre apart and the lowest of these lights shall, except where a towing light is required, be placed at a height of not less than 2 metres above the gunwale;

   (iii) when three lights are carried they shall be equally spaced.

(j) The lower of the two all-round lights prescribed for a vessel when engaged in fishing shall be at a height above the sidelights not less than twice the distance between the two vertical lights.

(k) The forward anchor light prescribed in Rule 30 *(a)* (i), when two are carried, shall not be less than 4·5 metres above the after one. On a vessel of 50 metres or more in length this forward anchor light shall be placed at a height of not less than 6 metres above the hull.

3. *Horizontal positioning and spacing of lights*

(a) When two masthead lights are prescribed for a power-driven vessel, the horizontal distance between them shall not be less than one-half of the length of the vessel but need not be more than 100 metres. The forward light shall be placed not more than one-quarter of the length of the vessel from the stem.

(b) On a power-driven vessel of 20 metres or more in length the sidelights shall not be placed in front of the forward masthead lights. They shall be placed at or near the side of the vessel.

(c) When the lights prescribed in Rule 27 *(b)* (i) or Rule 28 are placed vertically between the forward masthead light(s) and the after masthead light(s) these all-round lights shall be placed at a horizontal distance of not less than 2 metres from the fore and aft centreline of the vessel in the athwartship direction.

(d) When only one masthead light is prescribed for a power driven vessel, this light shall be exhibited forward of amidships; except that a vessel of less than 20 metres in length need not exhibit this light forward of amidships but shall exhibit it as far forward as is practicable.

4. *Details of location of direction-indicating lights for fishing vessels, dredgers and vessels engaged in underwater operations*

(*a*) The light indicating the direction of the outlying gear from a vessel engaged in fishing as prescribed in Rule 26 (*c*) (ii) shall be placed at a horizontal distance of not less than 2 metres and not more than 6 metres away from the two all-round red and white lights. This light shall be placed not higher than the all-round white light prescribed in Rule 26 (*c*) (i) and not lower than the sidelights.

(*b*) The lights and shapes on a vessel engaged in dredging or underwater operations to indicate the obstructed side and/or the side on which it is safe to pass, as prescribed in Rule 27 (*d*) (i) and (ii), shall be placed at the maximum practical horizontal distance, but in no case less than 2 metres, from the lights or shapes prescribed in Rule 27 (*b*) (i) and (ii). In no case shall the upper of these lights or shapes be at a greater height than the lower of the three lights or shapes prescribed in Rule 27 (*b*) (i) and (ii).

5. *Screens for sidelights*

The sidelights of vessels of 20 metres or more in length shall be fitted with inboard screens painted matt black, and meeting the requirements of Section 9 of this Annex. On vessels of less than 20 metres in length the sidelights, if necessary to meet the requirements of Section 9 of this Annex, shall be fitted with inboard matt black screens. With a combined lantern, using a single vertical filament and a very narrow division between the green and red sections, external screens need not be fitted.

6. *Shapes*

(*a*) Shapes shall be black and of the following sizes:

  (i) a ball shall have a diameter of not less than 0·6 metre;

 (ii) a cone shall have a base diameter of not less than 0·6 metre and a height equal to its diameter;

(iii) a cylinder shall have a diameter of at least 0·6 metre and a height of twice its diameter;

(iv) a diamond shape shall consist of two cones as defined in (ii) above having a common base.

(*b*) The vertical distance between shapes shall be at least 1·5 metre.

(*c*) In a vessel of less than 20 metres in length shapes of lesser dimensions but commensurate with the size of the vessel may be used and the distance apart may be correspondingly reduced.

7. *Colour specification of lights*

The chromaticity of all navigation lights shall conform to the following standards, which lie within the boundaries of the area of the diagram specified for each colour by the International Commission on Illumination (CIE).

The boundaries of the area for each colour are given by indicating the corner co-ordinates, which are as follows:

(i) *White*

x	0·525	0·525	0·452	0·310	0·310	0·443
y	0·382	0·440	0·440	0·348	0·283	0·382

(ii) *Green*

x	0·028	0·009	0·300	0·203
y	0·385	0·723	0·511	0·356

(iii) *Red*

x	0·680	0·660	0·735	0·721
y	0·320	0·320	0·265	0·259

(iv) *Yellow*

x	0·612	0·618	0·575	0·575
y	0·382	0·382	0·425	0·406

8. *Intensity of lights*

(*a*) The minimum luminous intensity of lights shall be calculated by using the formula:

$$I = 3 \cdot 43 \times 10^6 \times T \times D^2 \times K^{-D}$$

where I is luminous intensity in candelas under service conditions,

  T   is threshold factor 2×10^{-7} lux,

  D   is range of visibility (luminous range) of the light in nautical miles,

  K   is atmospheric transmissivity.

  For prescribed lights the value of K shall be $0 \cdot 8$, corresponding to a meteorological visibility of approximately 13 nautical miles.

(*b*) A selection of figures derived from the formula is given in the following table:

Range of visibility (luminous range) of light in nautical miles	Luminous intensity of light in candelas for $K = 0 \cdot 8$
D	I
1	$0 \cdot 9$
2	$4 \cdot 3$
3	12
4	27
5	52
6	94

Note: The maximum luminous intensity of navigation lights should be limited to avoid undue glare. This shall not be achieved by a variable control of the luminous density.

9. Horizontal sectors

(a) (i) In the forward direction, sidelights as fitted on the vessel shall show the minimum required intensities. The intensities shall decrease to reach practical cut-off between 1 degree and 3 degrees outside the prescribed sectors.

(ii) For sternlights and masthead lights and at 22·5 degrees abaft the beam for sidelights, the minimum required intensities shall be maintained over the arc of the horizon up to 5 degrees within the limits of the sectors prescribed in Rule 21. From 5 degrees within the prescribed sectors the intensity may decrease by 50 per cent up to the prescribed limits; it shall decrease steadily to reach practical cut-off at not more than 5 degrees outside the prescribed sectors.

(b) (i) All-round lights shall be so located as not to be obscured by masts, top-masts or structures within angular sectors of more than 6 degrees, except anchor light prescribed in Rule 30 which need not be placed at an impracticable height above the hull.

(b) (ii) If it is impracticable to comply with paragraph (b)(i) of this section by exhibiting only one all-round light, two all-round lights shall be used suitably positioned or screened so that they appear, as far as practicable, as one light at a distance of one mile.

10. *Vertical sectors*

(a) The vertical sectors of electric lights, as fitted, with the exception of lights on sailing vessels underway shall ensure that:

(i) at least the required minimum intensity is maintained at all angles from 5 degrees above to 5 degrees below the horizontal;

(ii) at least 60 per cent of the required minimum intensity is maintained from 7·5 degrees above to 7.5 degrees below the horizontal.

(b) In the case of sailing vessels underway the vertical sectors of electric lights as fitted shall ensure that:

(i) at least the required minimum intensity is maintained at all angles from 5 degrees above to 5 degrees below the horizontal;

(ii) at least 50 per cent of the required minimum intensity is maintained from 25 degrees above to 25 degrees below the horizontal.

(c) In the case of lights other than electric these specifications shall be met as closely as possible.

11. *Intensity of non-electric lights*

Non-electric lights shall so far as practicable comply with the minimum intensities, as specified in the Table given in Section 8 of this Annex.

12. *Manoeuvring light*

Notwithstanding the provisions of paragraph 2 (*f*) of this Annex the manoeuvring light described in Rule 34 (*b*) shall be placed in the same fore and aft vertical plane as the masthead light or lights and, where practicable, at a minimum height of 2 metres vertically above the forward masthead light, provided that it shall be carried not less than 2 metres vertically above or below the after masthead light. On a vessel where only one masthead light is carried the manoeuvring light, if fitted, shall be carried where it can best be seen, not less than 2 metres vertically apart from the masthead light.

13. *High speed craft*

The masthead light of high speed craft with a length to breadth ratio of less than 3.0 may be placed at a height related to the breadth of the craft lower than that prescribed in paragraph 2(a)(i) of this annex, provided that the base angle of the isosceles triangles formed by the sidelights and masthead light, when seen in end elevation, is not less than 27°.

14. *Approval*

The construction of lights and shapes and the installation of lights on board the vessel shall be to the satisfaction of the appropriate authority of the State whose flag the vessel is entitled to fly.

ANNEX II

Additional signals for fishing vessels fishing in close proximity

1. *General*

The lights mentioned herein shall, if exhibited in pursuance of Rule 26 (*d*), be placed where they can best be seen. They shall be at least 0·9 metre apart but at a lower level than lights prescribed in Rule 26 (*b*) (i) and (*c*) (i). The lights shall be visible all round the horizon at a distance of at least 1 mile but at a lesser distance than the lights prescribed by these Rules for fishing vessels.

2. *Signals for trawlers*

(*a*) Vessels of 20 metres or more in length when engaged in trawling, whether using demersal or pelagic gear, shall exhibit:

- (i) when shooting their nets:
    two white lights in a vertical line;
- (ii) when hauling their nets:
    one white light over one red light in a vertical line;
- (iii) when the net has come fast upon an obstruction:
    two red lights in a vertical line.

(*b*) Each vessel of 20 metres or more in length engaged in pair trawling shall exhibit:

- (i) by night, a searchlight directed forward and in the direction of the other vessel of the pair;

(ii) when shooting or hauling their nets or when their nets have come fast upon an obstruction, the lights prescribed in 2 (*a*) above.

(*c*) A vessel of less than 20 metres in length engaged in trawling, whether using demersal or pelagic gear or engaged in pair trawling, may exhibit the lights prescribed in paragraphs (*a*) or (*b*) of this section, as appropriate.

3. *Signals for purse seiners*

Vessels engaged in fishing with purse seine gear may exhibit two yellow lights in a vertical line. These lights shall flash alternately every second and with equal light and occultation duration. These lights may be exhibited only when the vessel is hampered by its fishing gear.

ANNEX III

Technical details of sound signal appliances

1. *Whistles*

(*a*) *Frequencies and range of audibility*

The fundamental frequency of the signal shall lie within the range 70–700 Hz.

The range of audibility of the signal from a whistle shall be determined by those frequencies, which may include the fundamental and/or one or more higher frequencies, which lie within the range 180–700 Hz ($\pm$ 1 per cent) and which provide the sound pressure levels specified in paragraph 1 (*c*) below.

(b) *Limits of fundamental frequencies*

To ensure a wide variety of whistle characteristics, the fundamental frequency of a whistle shall be between the following limits:

   (i) 70–200 Hz, for a vessel 200 metres or more in length;
   (ii) 130–350 Hz, for a vessel 75 metres but less than 200 metres in length;
   (iii) 250–700 Hz, for a vessel less than 75 metres in length.

(c) *Sound signal intensity and range of audibility*

A whistle fitted in a vessel shall provide, in the direction of maximum intensity of the whistle and at a distance of 1 metre from it, a sound pressure level in at least one 1/3rd-octave band within the range of frequencies 180–700 Hz ($\pm$ 1 per cent) of not less than the appropriate figure given in the table below.

Length of vessel in metres		*1/3rd-octave band level at 1 metre in dB referred to $2 \times 10^{-5} \ N/m^2$*	*Audibility range in nautical miles*
200 or more	.. ..	143	2
75 but less than 200	..	138	1·5
20 but less than 75	..	130	1
Less than 20	.. ..	120	0·5

The range of audibility in the table above is for information and is approximately the range at which a whistle may be heard on its forward axis with 90 per cent probability in conditions of still air on board a vessel having average background noise level at the listening posts (taken to be 68 dB in the octave band centred on 250 Hz and 63 dB in the octave band centred on 500 Hz).

In practice the range at which a whistle may be heard is extremely variable and depends critically on weather conditions; the values given can be regarded as typical but under conditions of strong wind or high ambient noise level at the listening post the range may be much reduced.

(d) Directional properties

The sound pressure level of a directional whistle shall be not more than 4dB below the prescribed sound pressure level on the axis at any direction in the horizontal plane within ± 45 degrees of the axis. The sound pressure level at any other direction in the horizontal plane shall be not more than 10 dB below the prescribed sound pressure level on the axis, so that the range in any direction will be at least half the range on the forward axis. The sound pressure level shall be measured in that 1/3rd-octave band which determines the audibility range.

(e) Positioning of whistles

When a directional whistle is to be used as the only whistle on a vessel, it shall be installed with its maximum intensity directed straight ahead.

A whistle shall be placed as high as practicable on a vessel, in order to reduce interception of the emitted sound by obstructions and also to minimize hearing damage risk to personnel. The sound pressure level of the vessel's own signal at listening posts shall not exceed 110 dB (A) and so far as practicable should not exceed 100 dB (A).

(*f*) *Fitting of more than one whistle*
   If whistles are fitted at a distance apart of more than 100 metres, it shall be so arranged that they are not sounded simultaneously.

(g) *Combined whistle systems*
   If due to the presence of obstructions the sound field of a single whistle or of one of the whistles referred to in paragraph 1 (*f*) above is likely to have a zone of greatly reduced signal level, it is recommended that a combined whistle system be fitted so as to overcome this reduction. For the purposes of the Rules a combined whistle system is to be regarded as a single whistle. The whistles of a combined system shall be located at a distance apart of not more than 100 metres and arranged to be sounded simultaneously. The frequency of any one whistle shall differ from those of the others by at least 10 Hz.

2. *Bell or gong*
(*a*) *Intensity of signal*
   A bell or gong, or other device having similar sound characteristics shall produce a sound pressure level of not less than 110 dB at a distance of 1 metre from it.

(b) *Construction*

Bells and gongs shall be made of corrosion-resistant material and designed to give a clear tone. The diameter of the mouth of the bell shall be not less than 300 mm. for vessels of 20 metres or more in length, and shall be not less than 200 mm. for vessels of 12 metres or more but of less than 20 metres in length. Where practicable, a power-driven bell striker is recommended to ensure constant force but manual operation shall be possible. The mass of the striker shall be not less than 3 per cent of the mass of the bell.

3. *Approval*

The construction of sound signal appliances, their performance and their installation on board the vessel shall be to the satisfaction of the appropriate authority of the State whose flag the vessel is entitled to fly.

ANNEX IV

Distress signals

1. The following signals, used or exhibited either together or separately, indicate distress and need of assistance:

(a) a gun or other explosive signal fired at intervals of about a minute;

(b) a continuous sounding with any fog-signalling apparatus:

(c) rockets or shells, throwing red stars fired one at a time at short intervals;

(d) a signal made by radiotelegraphy or by any other signalling method consisting of the group $\cdots---\cdots$ (SOS) in the Morse Code;

(e) a signal sent by radiotelephony consisting of the spoken word "Mayday";

(f) the International Code Signal of distress indicated by N.C.;

(g) a signal consisting of a square flag having above or below it a ball or anything resembling a ball;

(h) flames on the vessel (as from a burning tar barrel, oil barrel, etc.);

(i) a rocket parachute flare or a hand flare showing a red light;

(j) a smoke signal giving off orange-coloured smoke;

(k) slowly and repeatedly raising and lowering arms outstretched to each side;

(l) the radiotelegraph alarm signal;

(m) the radiotelephone alarm signal;

(n) signals transmitted by emergency position-indicating radio beacons.

(o) approved signals transmitted by radiocommunication systems, including survival craft radar transponders.

2.   The use or exhibition of any of the foregoing signals except for the purpose of indicating distress and need of assistance and the use of other signals which may be confused with any of the above signals is prohibited.

3.   Attention is drawn to the relevant sections of the International Code of Signals, the Merchant Ship Search and Rescue Manual and the following signals:

(a)   a piece of orange-coloured canvas with either a black square and circle or other appropriate symbol (for identification from the air);

(b)   a dye marker.